YOUR KNOWLEDGE HAS VALUE

- We will publish your bachelor's and master's thesis, essays and papers

- Your own eBook and book - sold worldwide in all relevant shops

- Earn money with each sale

Upload your text at www.GRIN.com
and publish for free

Microbial Assessment of Tamarind Drink Sold in Tamale Metropolis

Mubarik Mahamud

Bibliographic information published by the German National Library:

The German National Library lists this publication in the National Bibliography; detailed bibliographic data are available on the Internet at http://dnb.dnb.de.

ISBN: 9783346999078
This book is also available as an ebook.

© GRIN Publishing GmbH
Trappentreustraße 1
80339 München

Print and binding: Books on Demand GmbH, Norderstedt, Germany
Printed on acid-free paper from responsible sources.

The present work has been carefully prepared. Nevertheless, authors and publishers do not incur liability for the correctness of information, notes, links and advice as well as any printing errors.

GRIN web shop: https://www.grin.com/document/1439098

UNIVERSITY OF HEALTH AND ALLIED SCIENCES, HO

SCHOOL OF ALLIED HEALTH SCIENCES

DEPARTMENT OF NUTRITION AND DIETETICS

MICROBIAL ASSESSMENT OF TAMARIND DRINK SOLD IN TAMALE METROPOLIS

A PROJECT PROPOSAL SUBMITTED TO THE SCHOOL OF ALLIED HEALTH
SCIENCES, UNIVERSITY OF HEALTH AND ALLIED SCIENCES, HO, IN
PARTIAL FULFILMENT OF THE REQUIREMENT FOR THE AWARD OF
BACHELOR OF DIETETICS

MARCH, 2020

ABSTRACT

BACKGROUND: Tamarind drink is a local non-alcoholic beverage mostly processed and sold in Ghana by people of the Dagomba ethnic descend. The drink is a sweet-sour beverage made from the pulp of the indigenous African tamarind tree (*Tamarindus indica*). It has since ancient times been used for the treatment of inflammations, fevers, and alcohol intoxication. The juice from tamarind is rich in vitamin B3, potassium and is known to have cathartic, astringent, refrigerant, and antiseptic effects,

AIM: This study seeks to assess the microbial quality of Tamarind drink sold within the Tamale metropolis.

METHODS: The study will be empirical, involving a total of 30 tamarind drink samples purchased from ten different vendors all over the Tamale metropolis. Using a range of general and specific microbiological media including nutrient agar, MacConkey agar, Potato Dextrose Agar, enumeration of the microbes will be carried out. Colonies will be tested for Gram reaction, and the biochemical assay will be used to confirm morphological characteristics of the colonies following microscopic examination at x400 and x1000 magnifications. Questionnaires will be administered.

EXPECTED OUTCOME: The types and levels of pathogens associated with tamarind drinks produced by vendors within the Tamale metropolis will be known, and public health awareness on recommendations for tamarind drinks produced in the Tamale metropolis will be given at the end of the study.

TABLE OF CONTENTS

CHAPTER ONE

1.0 BACKGROUND

1.1 Introduction

Tamarind fruit drink, known as "Poha Beer" in Dagbani, "tankwan beer" in hausa and "Pusa daam" in Dagaare is a local non-alcoholic beverage mostly processed and sold by people of the Dagomba ethnic descend. "Poha Beer" is mostly found in the northern part of Ghana. (Dongdem *et al.*, 2020). Tamarind drink is a sweet-sour beverage made by infusing the dried pulp of *Tamarindus indica* L. It has since ancient times been used for the treatment of inflammations, fevers, and alcohol intoxication. The juice from tamarind is rich in vitamin B3, potassium and is known to have cathartic, astringent, refrigerant, and antiseptic effects. The hygienic conditions under which these beverages are produced are variable because of their wide prevalence among diverse socioeconomic groups and their low prices (Yamani, 2019).

Tamarind (*Tamarindus indica* L.) is a leguminous tree that belongs to the family Fabaceae and is indigenous to tropical Africa. The tamarind usually attains a maximum crown height of 12 to 18 meters (39 to 59 ft) (Menezes *et al.*, 2016). The fruit is an indehiscent legume, sometimes called a pod, 12 to 15 cm (4.7 to 5.9 in) in length, with a hard, brown shell (D'Cruz and Bonsai, 2011). The fruit when ripe possesses a fleshy, juicy, acidulous pulp. It is mature when the flesh turns coloured brown or reddish-brown. Asian tamarinds have longer pods (containing six to 12 seeds), whereas the tamarinds of Africa and West Indian varieties have shorter pods (containing one to six seeds). The seeds are somewhat flattened, and a glossy brown. The fruit is best described as sweet and sour and is high in tartaric acid, sugar, B vitamins, and, unusually for a fruit, calcium, but inadequate in Vitamin A, and Manganese and 3.5mg of Vitamin C.

Tamarind fruit pulp has a sweet acidic taste due to a combination of the high contents of tartaric acid and reducing sugars (De Caluwé *et al.*, 2009). The pulp is used for seasoning, in prepared foods, to flavor confections, curries, and sauces, and as a significant ingredient in juices and other beverages (De Caluwé *et al.*, 2009).

Tamarind is a famous tree in India (Shukla & Singh, 2020). It is also a good source of essential minerals, for example, Ca, Fe, Zn, Cu, Co, and Mn, but it may also contain metals that could potentially be toxic, such as Pb, Cd, Ni and Hg (Bagul *et al.*, 2015).

This study is critical because recently, the business of selling Tamarind drink has gained popularity bringing aboard more vendors. In order for these vendors to gain a competitive

1

market advantage over their ever-changing customers preferences, they adopt specific manufacturing, packaging and storage methods which may predispose the product to microbial contamination thereby warranting this investigation to ascertain their microbial quality and safety as suggested by Musah *et al.*, (2014). Again, the application of the results can assist food regulatory authorities in inspections while promoting consumer satisfaction.

1.2 Problem Statement

In Ghana, there is a rising popularity in the consumption of tamarind drink because consumers perceive locally produced drinks are healthier than their imported counterparts since they can quickly identify the ingredients they are familiar with in these drinks (Gamor *et al.*, 2015).

Tamarind drink is usually prepared fresh in, typical household and commercial outlets. Nonetheless, the method of preparation is unhygienic and inconvenient (Foo *et al.*, 2013). Besides, fresh unpasteurized tamarind juice has a long history of numerous outbreaks due to contamination of pathogenic microorganisms such as E. coli and Salmonella (*S. typhimurium*) (Daniyan & Muhammad, 2008). In recent times, there has been an increasing demand for tamarind drink due to its health benefits and uses in culinary which contribute to the production of commercial tamarind drink. However, the production of tamarind beverage is labor intensive, has low aesthetic value, poor hygiene practice, short product shelf life and no standardization of ingredients (Adeola & Aworh, 2013). This could be attributed to a lack of development and effort in tamarind drink processing. Previous finding by Chavasit *et al.* (2006) claimed that commercial tamarind beverages tested had no presence of coliform, *E.coli* and mould, but they were 100% contaminated by yeast. Thermal treatment is the conventional and widely used method in pasteurizing tamarind beverages. However, thermal treatment resulted in a change in flavor, appearance and especially nutritional content such as antioxidants in tamarind drink (Gabriel, 2015). This defeats the purpose of drinking tamarind drink, which is known to cure sore throat, fever and many other diseases. Tamarind fruits are also a potential source of microbes. The cloth used for extraction, if not well cleaned, could be another contributor. Preparation includes steps that incorporate exposure to sometimes unhygienic environmental conditions, and heat treatment is not used in the preparation. Citric acid reduces pH in tamarind drink. This shows that it is a potential source of microbial transmission due to the presence of various microbial contaminants found, hence Musah *et al.*, (2014) suggested the need for regular monitoring of the microbial quality of this drink. Hence, the microbial safety of locally prepared Tamarind drink in Tamale requires investigation and documentation.

1.3 Justification of the Study

The study is necessary because lately, grocery stores, school canteens, market places and restaurants are running the business of selling Tamarind drink. In order for these vendors to gain a competitive market advantage over their ever-changing customers preferences, they adopt specific manufacturing, packaging and storage methods which may predispose the product to microbial contamination thereby warranting this investigation to ascertain their microbial quality and safety as suggested by Musah *et al.*, (2014). Furthermore, the application of the results can assist food regulatory authorities in inspections while promoting consumer satisfaction.

1.4 Research Questions

1. What is the bacterial load in the tamarind drinks produced and sold in Tamale metropolis?
2. What types of pathogens can be found in the tamarind drinks produced and sold in Tamale metropolis.
3. What are the hygienic practices adopted by vendors of the tamarind drinks produced and sold in Tamale metropolis?

1.5 Objectives of the study

1.5.1 General objective

This study seeks to assess the microbial quality of Tamarind drink sold within the Tamale metropolis.

1.5.2 Specific objectives

- To investigate the production process of Tamarind drink sold in Tamale metropolis

- To assess the microbial load and species in sample positives sold in Tamale metropolis

- To determine the limits of the safety of microbes in the Tamarind drink sold in Tamale metropolis

- To investigate the hygienic practices of vendors of the beverage sold in Tamale metropolis

CHAPTER TWO

2.0 LITERATURE REVIEW

2.1 Introduction to Literature Review

This chapter reviews microbial assessment of Tamarind drink obtained from the Tamarind drink which will include; the classifications, the chemical composition, the pulp, the production and cultivation, the uses of the Tamarind products, and the functional properties.

2.2 Classification

The genus *Tamarindus* is a monotypic genus containing the sole species *T. indicus* and belongs to the sub-family *Caesalpinioideae* of the family *Fabaceae.* The most valuable part is the pod (also called the fruit). Pods are 7.5–20 cm long, 2.5 cm broad and 1 cm thick, more or less constricted between the seeds, slightly curved, brownish-ash coloured, scurfy. The outermost covering of the pod is fragile and easily separable (Chimsah *et al.*, 2020).

2.3 Chemical composition

The tamarind fruit consists mainly of pulp and seeds. The fruit, both ripe and dry, contains mainly tartaric acid, reducing sugars, pectin, tannin, fibre and cellulose. The whole seeds also contain protein, fat, sugars and carbohydrates. Both pulp and seeds are good sources of potassium, calcium and phosphorous and contain other minerals like sodium, zinc and iron (Soong & Barlow, 2004). The various components of tamarind are detailed in the following sections.

2.3.1 Pulp

The pulp of tamarind has a high magnesium content (25.6–30.2 mg/100 g), as is sodium (23.8–28.9 mg/100 g), whereas copper (0.8–1.2 mg/100 g) and zinc (0.8– 0.9 mg/100 g) are low. It also excels in riboflavin and is a good source of thiamin and niacin, but is deficient in vitamin A and vitamin C (De Caluwé *et al.*, 2009). The major volatile constituents of tamarind were reported by Soong & Barlow (2004). A review on traditional uses, phytochemistry and pharmacology of tamarind has been published by (De Caluwé *et al.*, 2009).The most outstanding characteristic of the tamarind fruit is that it is one of the most acidic of all fruits, because of its tartaric acid content which imparts the sour taste and outweighs the high total sugar content. Soong & Barlow (2004) reported that significant components of the volatiles were 2-phenyl acetaldehyde with a fruity and honey-like odour, 2-furfuryl with a caramel-like flavour and hexadecanoic acid and limonene having a citrus flavour. Volatile components of

4

tamarind fruits were isolated by simultaneous steam distillation/solvent extraction as well (Eduardo *et al.*, 2001).

2.4 Production and Cultivation

At present, tamarind is cultivated in 54 countries of the world: 18 in its native range, including central African countries, and 36 other countries including India and Thailand where it was introduced, and it has become naturalized in several regions. In the American continent, commercial plantations were reported in Belize, Central American countries and in north Brazil, The major producing countries are Brazil (Ekpong *et al.*, 2016), Bahamas, Costa Rica, Bangladesh, Cuba, Burma, Egypt, Cambodia, Guatemala, Dominican, Republic, India, Fiji, Indonesia, Gambia, Mexico, Kenya, Nicaragua, Pakistan, Puerto Rico, Senegal, Philippines, Tanzania, Sri Lanka, Vietnam, Thailand, Zambia, Venezuela and Zanzibar. However, tamarind is grown as a significant plantation only in a few countries such as India and Thailand. India is the world's largest producer of tamarind products.

2.5 Main Uses of Tamarind Products

2.5.1 Pulp

Tamarind is used in India, mainly in the form of pulp. The fruit pulp is the primary agent for souring curries, sauces, chutneys and certain beverages throughout the more significant part of India. In India, the immature green pods are often eaten by children and adults dipped in salt as a snack. It is also used in India to make 'tamarind fish', a seafood pickle, which is considered a great delicacy. In Egypt, tamarind is used to make a sour drink during the summer period, and it is also added to a similar lemon flavoured drink, popular in the Middle East. It is also used for this purpose in Mexico, where the drink is known as well as agua fresca (refreshing water) or agua de tamarindo, which is sometimes turned into frozen fruit ices. Mexicans also use tamarind as a snack, dried, salted or candied (e.g. Pulparindo). In the Philippines, it is also used to make sweets, but the leaves of the plant are also utilized in the recipe for the famous sinigang soup. In Guadeloupe, the fruit is used to make jam and syrup, whilst in northern Nigeria, tamarind is used during breakfast, as it is added to the traditional porridge known as pap or kunun tsamiya. Tamarind is also widely used in sauces to give a sour flavour, for example in the famous pad thai from Thailand, or in gravy for assam fish in Singapore and Malaya.

2.5.2 Tamarind Seeds

The seed of tamarind is used as a raw material in the manufacture of Tripotassium Phosphate (TKP), polysaccharide, adhesive, oil and tannin. Tamarind seed used to be an underutilized by-product of the tamarind pulp industry. Purified, refined tamarind xyloglucan is produced in Japan and is permitted as a thickening, stabilizing and gelling agent in the food, cosmetic and pharmaceutical industries (Yoko & Katsuyoshi, 2005). It is reported to possess properties like high viscosity, broad pH tolerance and adhesivity. Tamarind xyloglucan imparted more viscous, liquid-like rheological properties and heat stability to gelatinized tapioca starch/xyloglucan mixtures (Pongsawatmanit *et al.*, 2006).

2.5.3 Other Uses

Tamarind fruits and other extracts from the tree have several reported miscellaneous applications which are still in widespread use. The fruit pulp is used as a fixative with turmeric (Curcuma longa) and annatto (Bixa orellana) in dyeing, and it also serves to coagulate rubber latex and is used for ethanol production (Menon et al., 2010). Tamarind brown, the natural food colour from tamarind, is widely used in Japan as a food colourant (Anon., 2000). Kaur et al., (2010) reported that the red pigment (anthocyanin) from the half-matured red variety tamarind could be used to impart a natural and attractive red colour to curries, jam, jelly, etc. and that there is ample scope for red tamarind to be used as a source of natural red food colourant in the near future

2.6 Functional Properties

2.6.1 Medicinal Uses of Tamarind

The medicinal value of tamarind is mentioned in ancient Sanskrit literature. Havinga *et al.* (2010) extensively reviewed the ethnopharmacology of T. indica in the African context and suggested differences in the ways tamarind is used in local medicine in different parts of Africa. Anon. (2008) also detailed medicinal uses for tamarind in Africa which include as an anthelminthic (expels worms), antimicrobial, antiseptic, antiviral, sunscreen and astringent and to promote wound healing in the following conditions: asthma, bacterial skin infections, boils, chest pain, cholesterol metabolism disorders, colds, colic, conjunctivitis, constipation (chronic or acute), diabetes, diarrhoea, dry eyes, dysentery, eye inflammation, fever, gallbladder disorders, gastrointestinal disorders, gingivitis, haemorrhoids, indigestion, jaundice, keratitis, leprosy, liver disorders, nausea and vomiting (pregnancy-related), saliva production, skin disinfection/sterilization, sore throat, sores, sprains, swelling (joints) and urinary stones. It was suggested by Sadik (2010) that consumption of adequate amounts of 'poha beer' a popular tamarind fruit drinks of Northern Ghana in Africa, could help reduce the prevalence of iron deficiency anaemia. This was based on its vitamin C content which enhances the bioavailability of non-haem iron. Tamarind fruit is commonly used throughout Southeast Asia as a poultice applied to the foreheads of fever sufferers (Doughari, 2006). In traditional Thai medicine, the fruit of the tamarind is used as a digestive aid, carminative, laxative, expectorant and blood tonic (Komutarin *et al.*, 2003).

2.6.2 Antioxidant Activity

Several reports of antioxidant activity in tamarind indicate that fruits contain biologically essential mineral elements and have high antioxidant capacity associated with high phenolic content that can be considered beneficial to human health. Extraction of antioxidant compounds from the seed coat of sweet Thai tamarind was reported by Lourith & Kanlayavattanakul (2009) suggested that tamarind seed coat, a by-product of the tamarind gum industry, could be used as a safe and low-cost source of antioxidants, although other herbals could be more effective. Soong & Barlow (2004) reported that seeds of tamarind have higher antioxidant activity than that of the pulp. Martinello *et al.*, (2006) observed that the fruit pulp extract of T. indica, when administered at a concentration of 5 % to hypercholesteraemic, hamsters led to a decrease in total serum cholesterol and an increase in HDL, indicating its potential in diminishing the risk of atherosclerosis in humans. Sudjaroen *et al.* (2005) conducted a quantitative analysis of polyphenolic compounds in tamarind seeds and pericarp by analytical high-performance liquid chromatography. The yields of total phenolic compounds were 6.54 and 2.82 g/kg (dry weight) in the seeds and pericarp, respectively

2.6.3 Antimicrobial Properties

The fruits of tamarind are reported to have antifungal and antibacterial properties (Doughari, 2006). It is reported to be a potent fungicidal agent to cultures of Aspergillus niger and Candida albicans. Investigations by Daniyan & Muhammad (2008) revealed antimicrobial properties of tamarind against Escherichia coli, Klebsiella pneumoniae, Salmonella paratyphi, Pseudomonas aeruginosa and Staphylococcus aureus which are aetiological agents in urinary tract infections (UTI), wounds, pneumonia and paratyphoid fever. Extracts from tamarind fruit pulp have also shown molluscicidal activity against Bulius trancatus snails. Triterpenoids, phenols and alkaloids in tamarind extracts are being looked at for their use in controlling pests and diseases, e.g. control of citrus canker in Thailand (Leksomboon *et al.*, 2001). Tamarind pulp extracts screened for their antimicrobial activities exhibited higher activity against S. typhimurium and S. aureus and lower activity against A. niger (Jadhav *et al.*, 2019).

CHAPTER THREE

3.0 METHODOLOGY

3.1 Study Design

The study will be done in two phases; (1) questionnaires will be administered to the vendor to ascertain the hygienic practices they undertake, (2) samples of Tamarind drink will be collected from the different vendors at different locations in the Tamale Metropolis and taken to the lab for analyses to determine the microbial quality of the locally prepared drink. Only Tamarind drink in re-used and unused plastic bottles and rubbers that have been refrigerated will be sampled and analysed.

3.2 Study Site

The study setting is Tamale, the regional capital of the Northern Region of Ghana located between latitude 9° 16' and 9° 34' North and longitude 0° 36' and 0° 57' West in the Savannah wood lot. Climate is tropical continental with temperatures between 21O in the night and 32O C (Yakubu *et al.*, 2014). Tamale Metropolitan Area comprises two Sub-Metropolitan Areas, Tamale Central and Tamale South and has a land size of 646.9 sq. Km, a population of 223,252, with 111,109 (49.7%) being male and 112,143 (50.2%) being female (PHC,2010). The residents are predominantly Dagomba with other ethnic groups from other parts of the country. The predominant religion is Muslim

3.3 Study Population

The study population will include all vendors that sell Tamarind drink in the Tamale Metropolis.

3.4 Sample Size

A total of 10 vendors would be considered for the study and their tamarind drinks collected for the analyses.

3.5 Inclusion Criteria

All vendors who are in the business of selling tamarind drink within the study site will be included.

3.6 Exclusion criteria

Tamarind drinks produced and sold outside of the study site will be excluded.

3.7 Sampling technique

A convenient sampling technique will be used to obtain the samples.

3.8 Sampling and data collection

Survey data will be collected with the aid of a prepared questionnaire. Samples will be taken in the already made bottles and rubbers and then placed in labelled coolers.

3.9 Sample preparation

Under aseptic conditions, one millilitre (1 ml) of each sample collected will be pipetted into sterile sample bottles containing 9 ml of distilled water and mixed at 120 rev/min for 5-10 min on an orbital shaker (Gallenkamp, England). Ten serial dilutions will be prepared (10^{-1} – 10^{-5}) and aliquots of 1 ml of these dilutions will be used for analyses.

3.10 Bacteriological analyses

Bacteriological analyses will be done to estimate and identify pathogenic bacteria according to standard protocols for the number of heterotrophic bacteria, total coliforms, *staphylococcus aureus, Salmonella,* and *E. coli* species. The analyses will be done by serial dilution technique as in Aboagye *et al.* (2020) and plated on nutrient Agar, MacConkey Agar, Xylose lysine Agar and Mannitol Agar and then incubated at 37°C for 48 hours using the final dilutions.

Total plate count (TPC) of aerobic, mesophilic and micro-organisms colonies will be done after 48 hours and bacteria that appeared will be identified by morphological characterisation of their colonies. This means counts of duplicates will be calculated in colony-forming units per millilitre (CFU/ml).

3.11 Enumeration and identification of Fungi

This will be carried out according to the procedures used by Aboagye *et al.* (2020): one gram of each test sample will be added to 90ml of sterile distilled water and agitated vigorously and used as the stock solution. The samples will be serially diluted 10^{-1} up to 10^{-3} and then plated on Oxytetracycline Glucose Yeast Extract Agar (OGYEA) and Sabourand Dextrose Agar

(SDA) both for culturing fungi. It will be followed by incubation at 37°C for 5-7 days. After a week, moulds and yeast that may appear will be identified by their culture and morphological characteristics using standard identification protocols.

Enumeration will be done by colony counter (STAR 8500 Funke Gerber, Germany).

3.12 Determination of pH and temperature

The pH of samples will be determined directly with the aid of a bench pH meter (Jenway 3510, United Kingdom) after calibration using standard buffers 4.0 and 7.0 pH.

3.13 Data Handling

The collected samples would be immediately stored in coolers filled with ice at the point of collection and quickly transported to the University of Health and Allied Sciences microbiology laboratory to be stored in the freezer in order to prevent spoilage or any additional microbial growth on the samples collected.

3.14 Statistical Analyses

Analyses of samples will be carried out in triplicates. In each case, a mean value and standard deviation will be calculated. Analysis of variance (ANOVA) would also be performed, and the Duncan Multiple Range Test would be used to separate mean values at the 5% probability level. Results obtained from the experiment will be entered into a database and analysed statistically using Statistical Package for Social Sciences (SPSS) version 22.0 statistical software for windows. The relevant information would be represented in a standard form by way of tables, figures, frequencies and percentages for analysis and interpretation of data.

3.15 Ethical clearance

Ethical clearance will be obtained from the University of Health and Allied Sciences Ethical Review Board (UHAS, ERB). Findings will be for academic purposes and to generate Hazard Analysis of Critical Control Point plan to reduce the microbial load of tamarind drink.

REFERENCES

Aboagye, G., Gbolonyo-cass, S., & Korley, N. (2020). Microbial evaluation and some proposed good manufacturing practices of locally prepare d malted corn drink (" asaana ") and Hibiscus sabdarifa calyxes extract (" sobolo ") beverages sold at a university cafeteria in Ghana. *Scientific African*, *8*, e00330. https://doi.org/10.1016/j.sciaf.2020.e00330

Adeola, A. A., & Aworh, O. C. (2013). *Effects of sodium benzoate on storage stability of previously improved beverage from tamarind (Tamarindus indica L .).* 17–27. https://doi.org/10.1002/fsn3.78

Bagul, M., Sonawane, S. K., & Arya, S. S. (2015). *Tamarind seeds: chemistry, technology, applications and health benefits: A review. 34*(3).

Chimsah, F. A., Nyarko, G., & Abubakari, A.-H. (2020). A review of explored uses and study of nutritional potential of tamarind (Tamarindus indica L.) in Northern Ghana. *African Journal of Food Science*, *14*(9), 285–294. https://doi.org/10.5897/ajfs2018.1744

Daniyan, S. Y., & Muhammad, H. B. (2008). Evaluation of the antimicrobial activities and phytochemical properties of extracts of Tamaridus indica against some diseases causing bacteria. *African Journal of Biotechnology*, *7*(14), 2451–2453. https://doi.org/10.5897/AJB08.337

De Caluwé, E., Halamová, K., & Van Damme, P. (2009). Baobab (Adansonia digitata L.): A review of traditional uses, phytochemistry and pharmacology. *ACS Symposium Series*, *1021*(1), 51–84. https://doi.org/10.1021/bk-2009-1021.ch004

Dongdem, J. T., Abugri, J., & Asakedola, C. A. (2020). *Chemical Composition and Microbial Contaminants of Poha Beer : A Local Nonalcoholic Beverage in the Bolgatanga Municipality , Ghana. 2020.*

Doughari, J. H. (2006). *Antimicrobial Activity of Tamarindus indica Linn. 5*(December), 597–603.

Eduardo, J. V.-C., Espinosa-Paredes, G., Beristain, C. I., & Romero-Tehuitzil, H. (2001). *Effect of foaming agents on the stability, rheological properties, drying kinetics and flavour retention of tamarind foam-mats* (pp. 537–598).

Ekpong, A., Phomkong, W., & Onsaard, E. (2016). The effects of maltodextrin as a drying aid and drying temperature on production of tamarind powder and consumer acceptance of the powder. *International Food Research Journal*, *23*(1), 300–308.

Foo, K. Y., Lee, L. K., & Hameed, B. H. (2013). Preparation of tamarind fruit seed activated carbon by microwave heating for the adsorptive treatment of landfill leachate: A laboratory column evaluation. *Bioresource Technology, 133,* 599–605. https://doi.org/10.1016/j.biortech.2013.01.097

Gabriel, Y. (2015). *The Unmanageable Consumer.*

Gamor, G., Akoto-Danso, E. K., Karg, H., & Chagomoka, T. (2015). *Traditional Recipes from the Northern Region of Ghana.*

Havinga, R. M., Hartl, A., Putscher, J., Prehsler, S., Buchmann, C., & Vogl, C. R. (2010). Tamarindus indica L. (Fabaceae): Patterns of use in traditional African medicine. *Journal of Ethnopharmacology, 127*(3), 573–588. https://doi.org/10.1016/j.jep.2009.11.028

Jadhav, U., Cavazza, A., Banerjee, K. K., Xie, H., O'Neill, N. K., Saenz-Vash, V., Herbert, Z., Madha, S., Orkin, S. H., Zhai, H., & Shivdasani, R. A. (2019). Extensive Recovery of Embryonic Enhancer and Gene Memory Stored in Hypomethylated Enhancer DNA. *Molecular Cell, 74*(3), 542-554.e5. https://doi.org/10.1016/j.molcel.2019.02.024

Kaur, G., Jain, S., & Tiwary, A. K. (2010). Chitosan-carboxymethyl tamarind kernel powder interpolymer complexation: Investigations for colon drug delivery. *Scientia Pharmaceutica, 78*(1), 57–78. https://doi.org/10.3797/scipharm.0908-10

Komutarin, T., Azadi, S., Butterworth, D., Keil, B., Chitosomboon, M., Suttajit, M., & Meade, B. J. (2003). Extracts of the seed coat of Tamarindus Indica inhibits nitric oxide production by murine macrophages in vitro and in vivo. *Food and Chemical Technology, 42*(2004), 649–658.

Leksomboon, C., Thaveechai, N., & Kositratana, W. (2001). *Potential of Plant Extracts for Controlling Citrus Canker of Lime. 396,* 392–396.

Lourith, N., & Kanlayavattanakul, M. (2009). Natural surfactants used in cosmetics: Glycolipids. *International Journal of Cosmetic Science, 31*(4), 255–261. https://doi.org/10.1111/j.1468-2494.2009.00493.x

Martinello, F., Soares, S. M., Franco, J. J., Santos, A. C., Sugohara, A., Garcia, S. B., Curti, C., & Uyemura, S. A. (2006). Hypolipemic and antioxidant activities from Tamarindus indica L. pulp fruit extract in hypercholesterolemic hamsters. *Food and Chemical Toxicology, 44*(6), 810–818. https://doi.org/10.1016/j.fct.2005.10.011

Menezes, A. P. P., Trevisan, S. C. C., Barbalho, S. M., & Guiguer, E. L. (2016). Tamarindus indica L. A plant with multiple medicinal purposes. *Journal of Pharmacognosy and Phytochemistry*, *5*(53), 50–54. http://www.phytojournal.com/archives/2016/vol5issue3/PartA/5-3-7-335.pdf

Menon, V., Prakash, G., Prabhune, A., & Rao, M. (2010). Biocatalytic approach for the utilization of hemicellulose for ethanol production from agricultural residue using thermostable xylanase and thermotolerant yeast. *Bioresource Technology*, *101*(14), 5366–5373. https://doi.org/10.1016/j.biortech.2010.01.150

Musah, B. O., Nii-trebi, N. I., Nwabugo, M. A., & Asmah, R. H. (2014). Microbial Quality of Locally Prepared Hibiscus Tea in Accra. *IOSR Journal of Environmental Science, Toxicology and Food Technology*, *8*(11), 23–27. https://doi.org/10.9790/2402-081122327

Rungnaphar Pongsawatmanit, Theeranan Temsiripong, Shinya Ikeda, K. N. (2006). Influence of tamarind seed xyloglucan on rheological properties and thermal stability of tapioca starch. *Journal of Food Engineering*, *77*(1), 41–50. https://doi.org/https://doi.org/10.1016/j.jfoodeng.2005.06.017

Sadik, H. A. (2010). The nutritional value of "Poha Beer" (tamarind fruit drink) and its social usage in tamale metropolis. *Pakistan Journal of Nutrition*, *9*(8), 797–805.

Shukla, A. K., & Singh, J. (2020). Studies on physico-chemical evaluation of tamarind (Tamarindus indica L.) genotypes prevailing in bastar region of Chhattisgarh on macro nutrient status of tamarind seed. *International Journal of Chemical Studies*, *8*(2), 2317–2320. https://doi.org/10.22271/chemi.2020.v8.i2ai.9094

Soong, Y. Y., & Barlow, P. J. (2004). Antioxidant activity and phenolic content of selected fruit seeds. *Food Chemistry*, *88*(3), 411–417. https://doi.org/10.1016/j.foodchem.2004.02.003

Visith Chavasit, Supaporn Kunhawattana, W. J. (2006). Production and contamination of pasteurized beverages packed in sealed plastic containers in Thailand and potential preventive measures. *Food Control*, *17*(8), 622–630.

Yamani, M. I. (2019). *Microbiological Quality of Sous and Tamarind , Traditional Drinks Consumed Microbiological Quality of Sous and Tamarind , Traditional Drinks Consumed in Jordan. March.* https://doi.org/10.4315/0362-028X-68.4.773

Yoko, N., & Katsuyoshi, N. (2005). Gelation and gel properties of polysaccharides gellan gum and tamarind xyloglucan. *Biological Macromolecules*, *5*(3), 47–52.

WORK PLAN

Activity	Time Frame
Proposal write-up and submission	14th December 2020 – 5th March 2021
Administration of questionnaires and sample collection and transportation	5th March 2021 – 14th April 2021
Laboratory analyses	14 April 2021– 13th May 2021
Data entry and analysis	14th April 2021 – 13th May 2021
Dissertation write-up	13th May 2021 – 10th June 2021
Submission of dissertation	10th June – 13th June 2021

BUDGET

Item/ Activity	Unit price (GHc)	Total (GHc)
Lab reagents	200	800
Printing of proposal	5	80
Questionnaires	2	20
Purchase of samples	2	60
Transportation		500
Miscellaneous		200
Printing dissertation	20	100
Total		**1760**